JUMBO COLORING BOOK

TUNDRA ECOSYSTEMS ARE TREELESS AREAS IN THE ARCTIC AND ON MOUNTAIN TOPS, WHERE IT'S COLD, WINDY, AND DRY.

Biographical Note
Tundra Ecosystems An Educational Coloring Book is a new work,
first published by Little Artist Studio in 2025.

International Standard Book Number
ISBN 979-8-9992504-1-4

www.littleartiststudio.org

EXPLORE THE BEAUTY AND POWER OF THE WORLD'S TUNDRA ECOSYSTEMS WITH THIS CAPTIVATING COLORING BOOK! FEATURING OVER 85 INTRICATELY DETAILED, FULL-PAGE ILLUSTRATIONS FROM AROUND THE GLOBE, THIS BOOK IS DESIGNED TO EDUCATE AND INSPIRE A DEEPER UNDERSTANDING OF OUR PLANET'S FRAGILE ENVIRONMENTS. PART OF LITTLE ARTIST STUDIO'S ACCLAIMED EDUCATIONAL COLORING SERIES, EACH IMAGE TELLS A COMPELLING STORY THAT SPARKS CURIOSITY AND CREATIVITY. WITH SINGLE-SIDED PAGES, ARTISTS OF ALL AGES CAN USE ANY COLORING MEDIUM AND EASILY DISPLAY THEIR FINISHED MASTERPIECES. PERFECT FOR NATURE LOVERS, EDUCATORS, AND CREATIVE MINDS ALIKE.

ARCTIC TUNDRA BIOME

A BIOME IS A BIG AREA OF LAND WITH CERTAIN PLANTS AND ANIMALS.

ARCTIC TUNDRA BIOME

EACH BIOME HAS MANY ECOSYSTEMS, WHICH ARE PLACES WHERE ANIMALS, PLANTS, WATER, AND LAND ALL WORK TOGETHER.

ARCTIC TUNDRA BIOME

THE ARCTIC TUNDRA BIOME INCLUDES LAND AREAS SUCH AS ALASKA, CANADA, RUSSIA, AND GREENLAND, AS WELL AS REGIONS BORDERING THE ARCTIC OCEAN AND NEARBY SEAS.

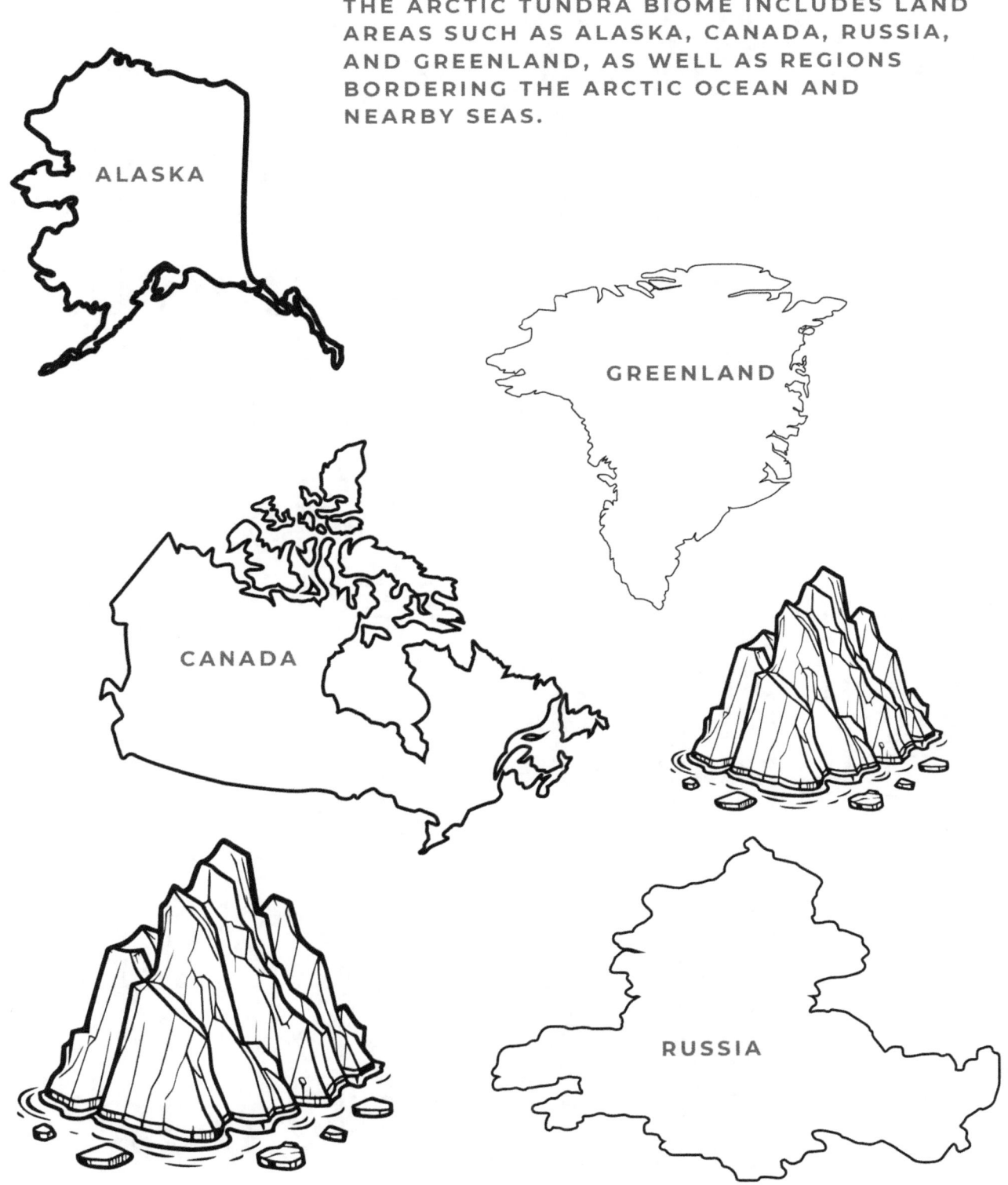

ALASKA

GREENLAND

CANADA

RUSSIA

ARCTIC TUNDRA BIOME

WHILE THE ARCTIC OCEAN ITSELF IS NOT PART OF THE TUNDRA BIOME, THE SEA ICE AND MARINE ECOSYSTEMS ALONG ITS EDGES CLOSELY INTERACT WITH THE TUNDRA LANDSCAPE.

ARCTIC TUNDRA

WHILE LYNX CAN BE FOUND IN TUNDRA AREAS WHEN PREY IS SCARCE IN THEIR USUAL FORESTED HABITATS, THEIR PRIMARY RANGE IS THE BOREAL FOREST (TAIGA) OF CANADA AND ALASKA.

LYNX

ARCTIC TUNDRA

NARWHALS ARE MARINE MAMMALS UNIQUELY ADAPTED TO COLD ENVIRONMENTS. THEY LIVE IN THE ARCTIC OCEAN, PARTICULARLY IN THE ICY WATERS NEAR THE ARCTIC TUNDRA.

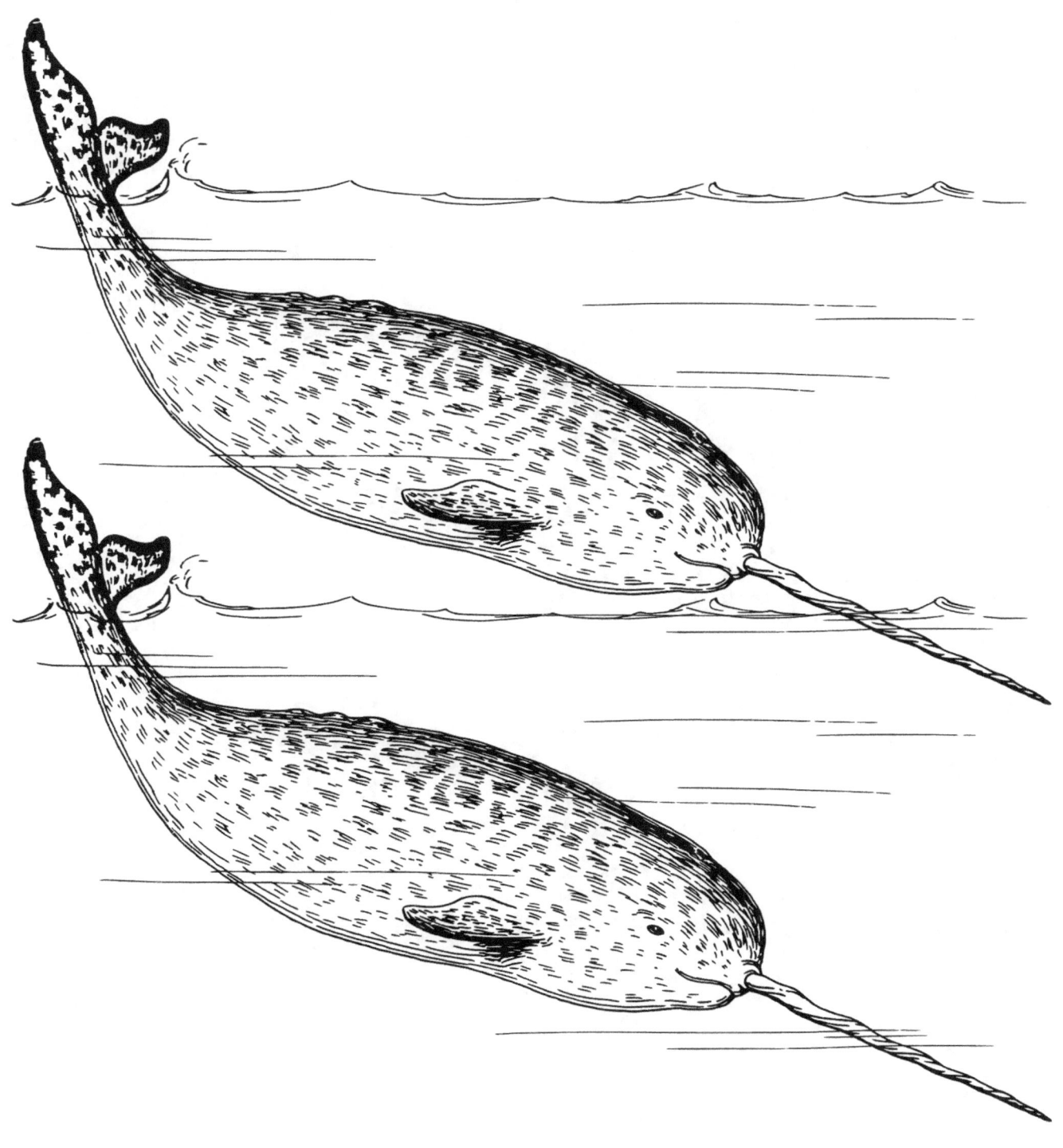

ARCTIC TUNDRA

THE ARCTIC TUNDRA STRETCHES DOWN TO
PLACES LIKE HUDSON BAY IN CANADA AND THE
TOP OF ICELAND. THERE IS ALSO ALPINE
TUNDRA, WHICH IS FOUND ON HIGH MOUNTAINS.

ARCTIC TUNDRA

ARCTIC TUNDRA

ARCTIC TUNDRA

ONE IMPORTANT THING ABOUT THE TUNDRA IS
THE PERMAFROST, WHICH MEANS THE GROUND
IS PERMANENTLY FROZEN.

ARCTIC TUNDRA

IN WINTER, ALMOST ALL THE GROUND IS
FROZEN. IN SUMMER, ONLY THE TOP LAYER OF
THE GROUND MELTS.

ARCTIC TUNDRA

PERMAFROST KEEPS PLANT ROOTS FROM GROWING DEEP AND STOPS TREES FROM GROWING.

ARCTIC TUNDRA

EVEN WITHOUT TREES, THE TUNDRA HOLDS A
LOT OF CARBON IN THE GROUND.

ARCTIC TUNDRA

PEAT IS MADE OF DEAD SPHAGNUM MOSS, AND
HUMUS IS ORGANIC MATTER-LIKE DEAD PLANTS
AND MOSS-THAT HAS SLOWLY BROKEN DOWN
OVER TIME.

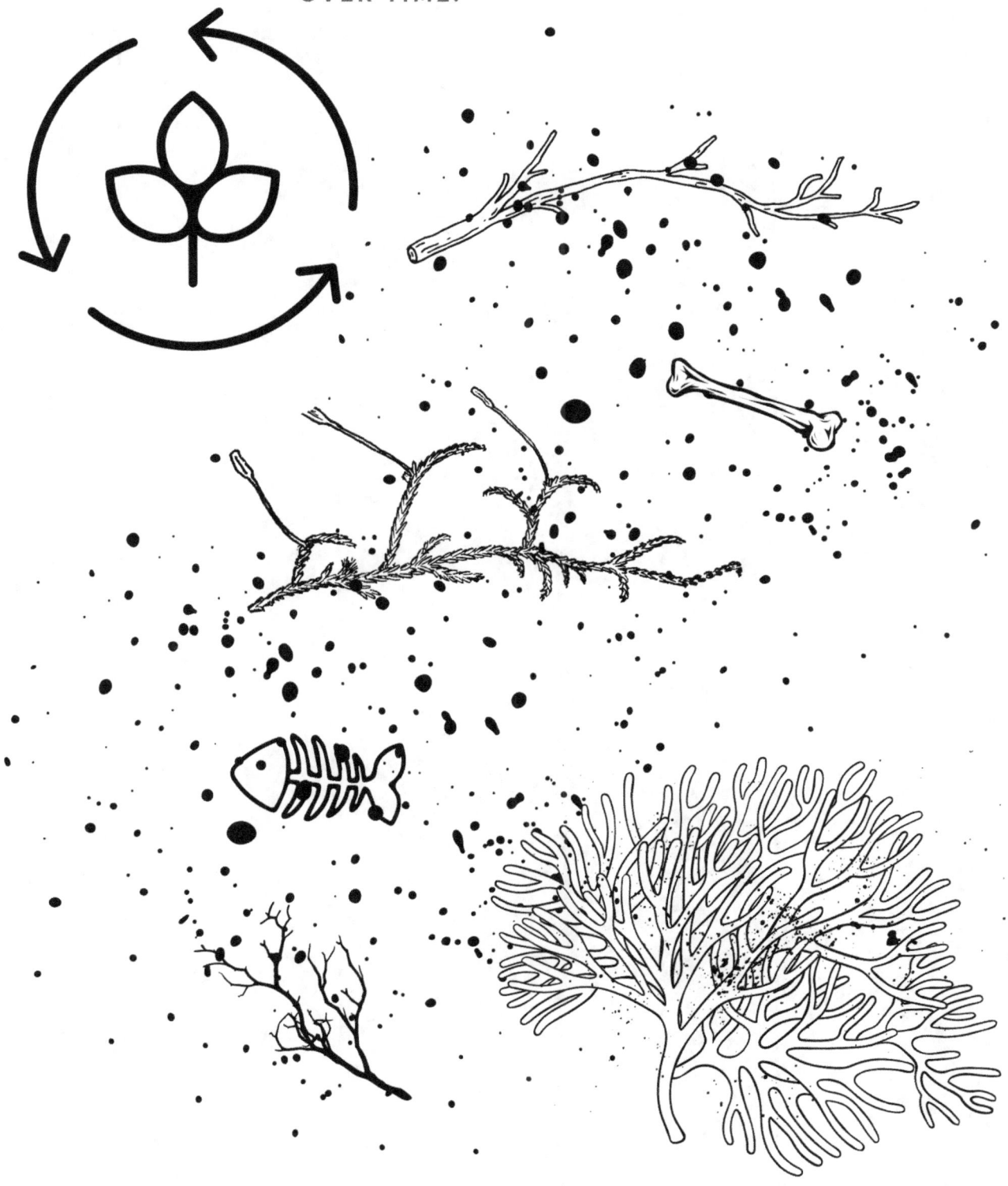

ANTARCTIC TUNDRA

ANTARCTIC TUNDRA IS FOUND IN ANTARCTICA
AND THE ISLANDS AROUND IT.

ANTARCTIC TUNDRA

ARCTIC COUNTRIES

THERE ARE EIGHT COUNTRIES IN
THE ARCTIC CIRCLE - CANADA,
UNITED STATES (ALASKA),RUSSIA,
NORWAY, FINLAND, SWEDEN,
ICELAND, DENMARK (GREENLAND).

ARCTIC COUNCIL

THE ARCTIC COUNCIL IS A
GROUP THAT HELPS ARCTIC
COUNTRIES WORK TOGETHER
AND SHARE IDEAS

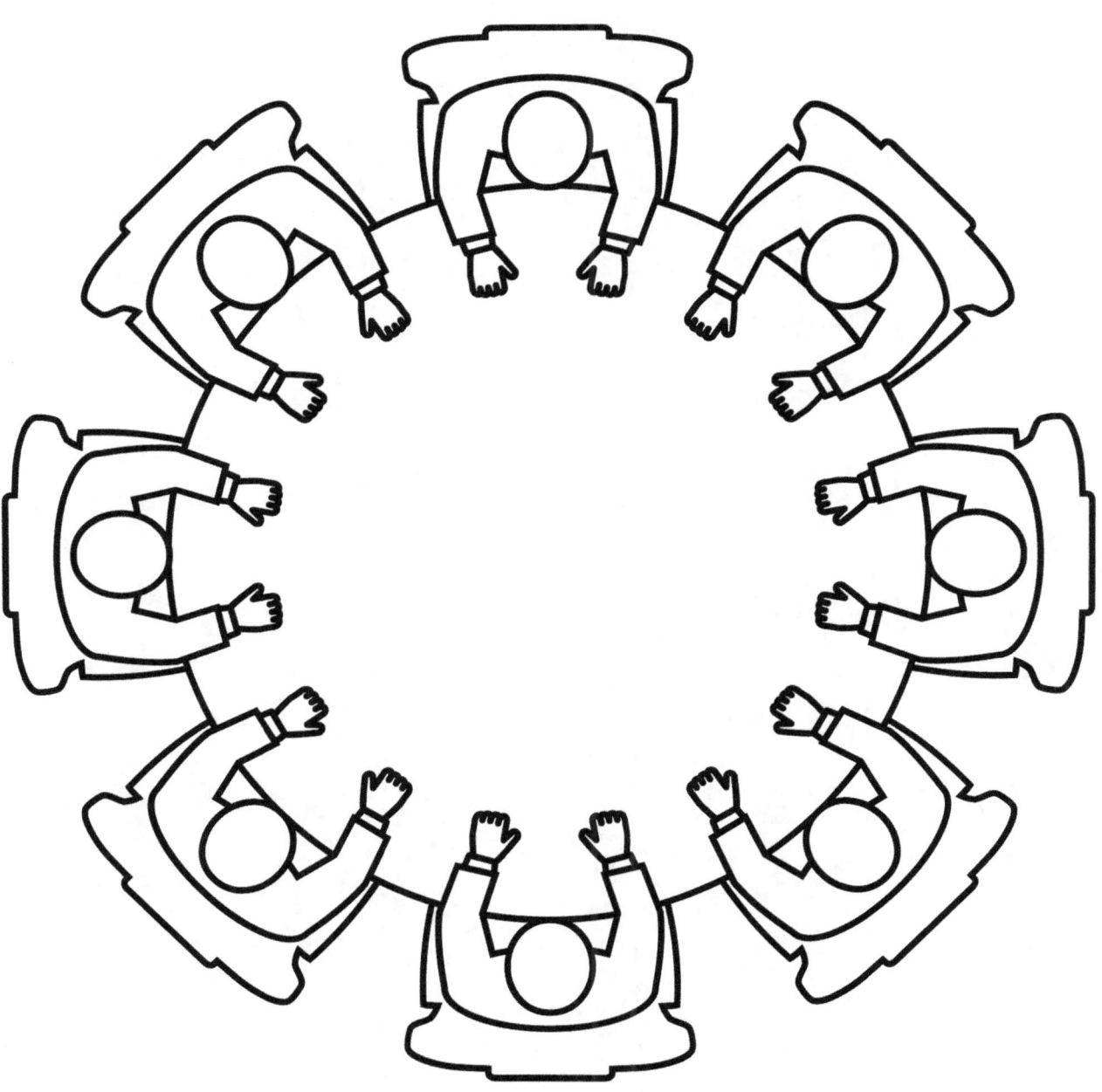

ARCTIC LANDSCAPES

GLACIERS ARE HUGE AND HEAVY, SO THEY CAN SLOWLY MOVE LIKE A FROZEN RIVER. SOME ARE SO OLD THEY'VE BEEN AROUND SINCE THE ICE AGE!

ARCTIC LANDSCAPES

LONG AGO, INUIT PEOPLE USED
IGLOOS AS TEMPORARY HOUSES IN
THE WINTER, ESPECIALLY WHEN
THEY WERE HUNTING.

ARCTIC LANDSCAPES

AN IGLOO IS A ROUND HOUSE
MADE OF SNOW BLOCKS.

ARCTIC LANDSCAPES

THE ARCTIC HAS SOME OF THE BIGGEST AND DEEPEST LAKES IN THE WORLD, LIKE GREAT BEAR LAKE, LAKE TAYMYR, AND GREAT SLAVE LAKE - WELL KNOWN DESTINATIONS FOR ARCTIC CHAR FISHING.

ARCTIC LANDSCAPES

GREENLAND HAS MANY COLD
LAKES, SOME COVERED WITH
ICE MOST OF THE YEAR.

PLANTS AND ANIMALS

MOST PLANTS IN THE TUNDRA ARE SOFT-
STEMMED AND DON'T HAVE HARD, WOODY
STEMS LIKE TREES OR BUSHES.

PLANTS AND ANIMALS

GRASSES AND MOSSES ARE SOME OF THE SOFT
PLANTS THAT GROW IN THE TUNDRA. EXAMPLES
INCLUDE REINDEER MOSS, LIVERWORTS, AND
LICHENS.

PLANTS AND ANIMALS

MOST TREES CAN'T GROW IN THE TUNDRA
BECAUSE IT'S TOO COLD. BUT WHITE BIRCH CAN
GROW THERE, AND MOUNTAIN BIRCH GROWS
NEARBY.

WHITE BIRCH

PLANTS AND ANIMALS

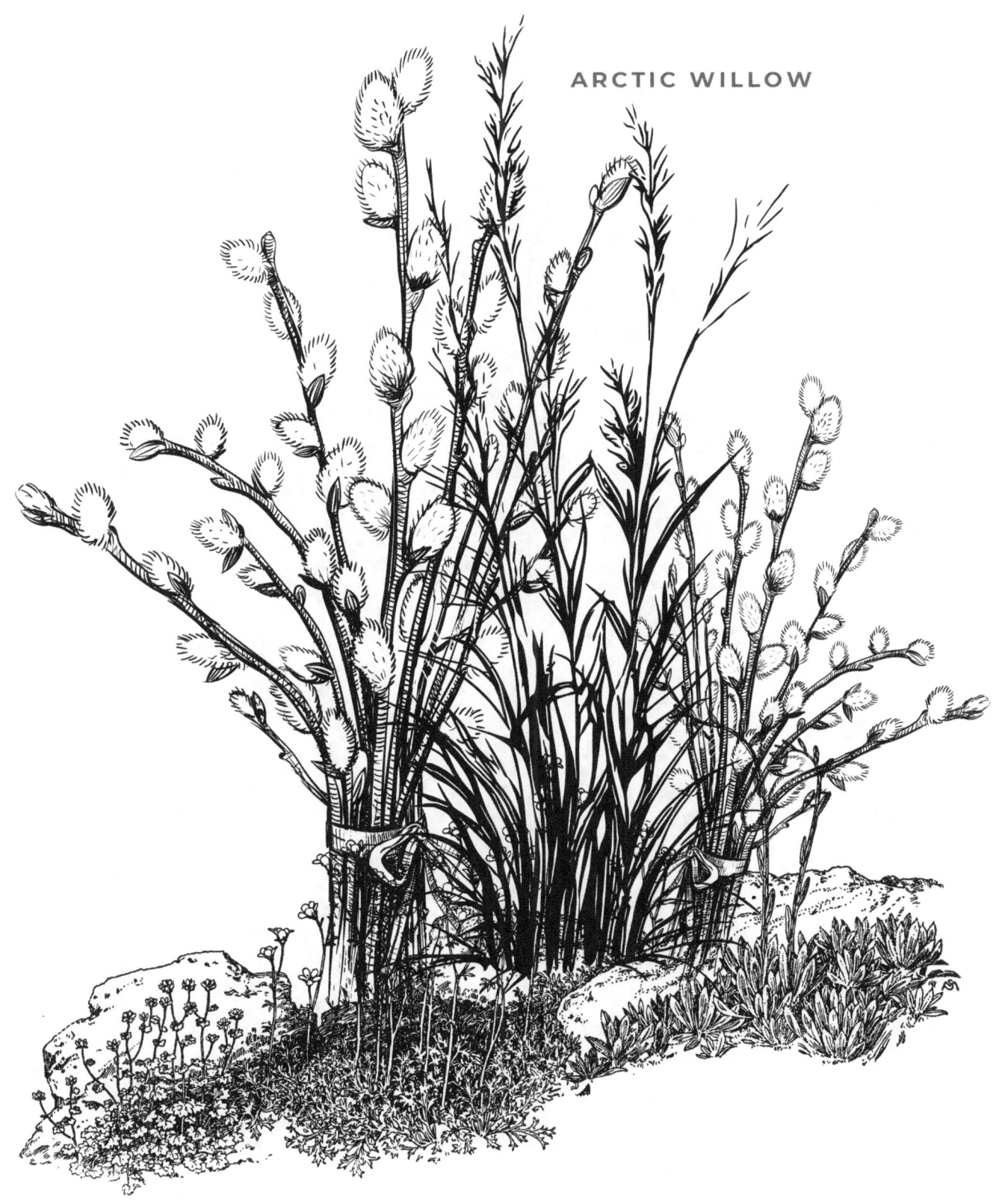

ARCTIC WILLOW

PLANTS AND ANIMALS

THE PURPLE SAXIFRAGE, IS A SMALL FLOWER THAT GROWS IN
COLD, ROCKY PLACES. IT HAS PURPLE OR PINK FLOWERS AND
BLOOMS EARLY IN THE SPRING, EVEN THROUGH THE SNOW!

PLANTS AND ANIMALS

COTTONGRASS IS A PLANT THAT GROWS IN
WET, COLD PLACES. IT HAS FLUFFY WHITE
PARTS THAT LOOK LIKE COTTON AND
HELPS PROTECT THE SOIL.

PLANTS AND ANIMALS

A DWARF SHRUB IS A SMALL, BUSHY PLANT THAT GROWS CLOSE TO THE GROUND. IT DOESN'T GROW VERY TALL AND IS OFTEN FOUND IN COLD OR WINDY PLACES LIKE THE ARCTIC.

PLANTS AND ANIMALS

LICHEN HELPS FEED SOME ANIMALS, LIKE REINDEER, AND IS VERY IMPORTANT IN THE ARCTIC ECOSYSTEM.

PLANTS AND ANIMALS

MANY LARGE MAMMALS, SUCH AS
CARIBOU, POLAR BEARS, ARCTIC
FOXES, AND MUSK OX, ARE FOUND IN
THIS BIOME.

PLANTS AND ANIMALS

THE ARCTIC FOX HAS A SPECIAL FUR
THAT KEEPS THEM WARM IN THE ICY
ARCTIC, MAKING THEIR BODY AS COZY
AS 104°F.

PLANTS AND ANIMALS

POLAR BEARS LIVE IN ONE OF THE
COLDEST PLACES ON EARTH AND STAY
WARM THANKS TO THEIR THICK FUR AND
A COZY LAYER OF FAT UNDERNEATH!

PLANTS AND ANIMALS

GRAY WOLVES

PLANTS AND ANIMALS

CARIBOU

PLANTS AND ANIMALS

MUSK OXEN

PLANTS AND ANIMALS

SNOW GEESE

PLANTS AND ANIMALS

GRIZZLY BEAR

WOLF SPIDER

PLANTS AND ANIMALS

A PIKA IS A SMALL, FURRY ANIMAL THAT
LIVES IN COLD, ROCKY PLACES AND
LOOKS LIKE A TINY RABBIT.

PLANTS AND ANIMALS

WALRUSES LIVE IN THE ARCTIC OCEAN
AND HAVE BIG TUSKS AND WHISKERS.

PLANTS AND ANIMALS

SNOWY OWL

PLANTS AND ANIMALS

THE ARCTIC SNAIL IS A SPECIAL KIND OF MARINE
MOLLUSK CALLED A PTEROPOD, ALSO KNOWN AS
A "SEA BUTTERFLY."

PLANTS AND ANIMALS

RED FOXES DO NOT HIBERNATE; THEY STAY
ACTIVE AND FIND WARM PLACES TO SLEEP IN
WINTER STORMS.

PLANTS AND ANIMALS

THE BAR-TAILED GODWIT (LIMOSA LAPPONICA), A MIGRATORY BIRD, HOLDS THE RECORD FOR THE LONGEST NON-STOP FLIGHT BY A BIRD.

PLANTS AND ANIMALS

PLANTS AND ANIMALS

LAPLAND LONGSPURS

PLANTS AND ANIMALS

THE AMERICAN MINK (MUSTELA VISON) IS AN
INVASIVE SPECIES IN SOME PARTS OF THE
ARCTIC AND CAN HARM NATIVE ANIMALS
AND ECOSYSTEMS.

INUIT

INUIT ARE 1 OF 3 RECOGNIZED INDIGENOUS PEOPLES IN CANADA, ALONG WITH FIRST NATIONS AND MÉTIS. THE WORD INUIT MEANS "THE PEOPLE".

INUIT

THE INUIT ARE INDIGENOUS PEOPLE WHO LIVE
IN THE ARCTIC. THE WORD INUIT MEANS "THE
PEOPLE" IN THEIR LANGUAGE, INUKTUT. ONE
PERSON IS CALLED AN INUK.

INUIT

MANY INUIT IN CANADA LIVE IN
INUIT NUNANGAT, WHICH MEANS
"THE PLACE WHERE INUIT LIVE."

INUKSHUK

AN INUKSHUK IS A STONE
FIGURE BUILT BY INUIT
PEOPLE IN THE ARCTIC.

INUKSHUK

INUKSHUK

INUKSHUK WAS USED TO SHOW DIRECTIONS, MARK SPECIAL PLACES, OR EVEN HELP WITH HUNTING BY GUIDING CARIBOU.

INUIT NUNANGAT

MANY INUIT IN CANADA LIVE IN INUIT NUNANGAT, WHICH MEANS "THE PLACE WHERE INUIT LIVE."

INUKTUT

INUKTUT IS SPOKEN THROUGHOUT
INUIT NUNANGAT, AND EACH REGION
HAS ITS OWN DIALECTS.

QAMUTIK

THE INUIT TRADITIONAL SLED USED FOR HUNTING IS CALLED A QAMUTIK (ALSO SPELLED QAMUTIIK)

TUNDRA LIFE

IN THE SUMMER, THE NORTH POLE HAS
SUNLIGHT ALL DAY LONG, WHICH IS WHY
THE ARCTIC IS CALLED THE 'LAND OF THE
MIDNIGHT SUN'

TUNDRA LIFE

IN SUMMER, TUNDRA FLOWERS BLOOM
QUICKLY DURING THE SHORT WARM
SEASON.

TUNDRA LIFE

ICE FISHING IN THE ARCTIC HAPPENS IN WINTER WHEN THE WATER IS FROZEN SOLID. IT GIVES PEOPLE FRESH FISH IN PLACES WHERE FARMING IS HARD.

TUNDRA LIFE

POLAR NIGHT ARE PERIODS WHEN THE SUN
DOESN'T RISE FOR MANY DAYS OR EVEN
MONTHS, DEPENDING ON HOW FAR NORTH YOU
ARE. IT STAYS DARK OR DIM ALL DAY, CREATING
A QUIET AND MYSTERIOUS FEELING.

TUNDRA LIFE

NORTHERN LIGHTS ARE BRIGHT, COLORFUL LIGHTS IN THE NIGHT SKY. THEY ARE CAUSED BY THE SUN AND ARE EASIEST TO SEE IN THE ARCTIC DURING DARK WINTER NIGHTS.

SIBERIAN HUSKY

TUNDRA LIFE

DOG SLEDDING IS A TRADITIONAL WAY TO TRAVEL OVER SNOW AND ICE USING SLED DOGS. IT'S STILL USED IN SOME PLACES AND IS ALSO A POPULAR ACTIVITY FOR TOURISTS.

TUNDRA LIFE

TUNDRA LIFE

ARCTIC ECONOMICS

ARCTIC ECONOMICS IS ABOUT HOW PEOPLE USE THE
LAND, OCEAN, AND RESOURCES IN THE COLD,
NORTHERN PART OF THE WORLD CALLED THE
ARCTIC.

ARCTIC ECONOMICS

FISHERIES IN THE ARCTIC ARE A MAJOR
SOURCE OF FOOD, JOBS, AND INCOME,
ESPECIALLY FOR GREENLAND, CANADA AND
RUSSIA.

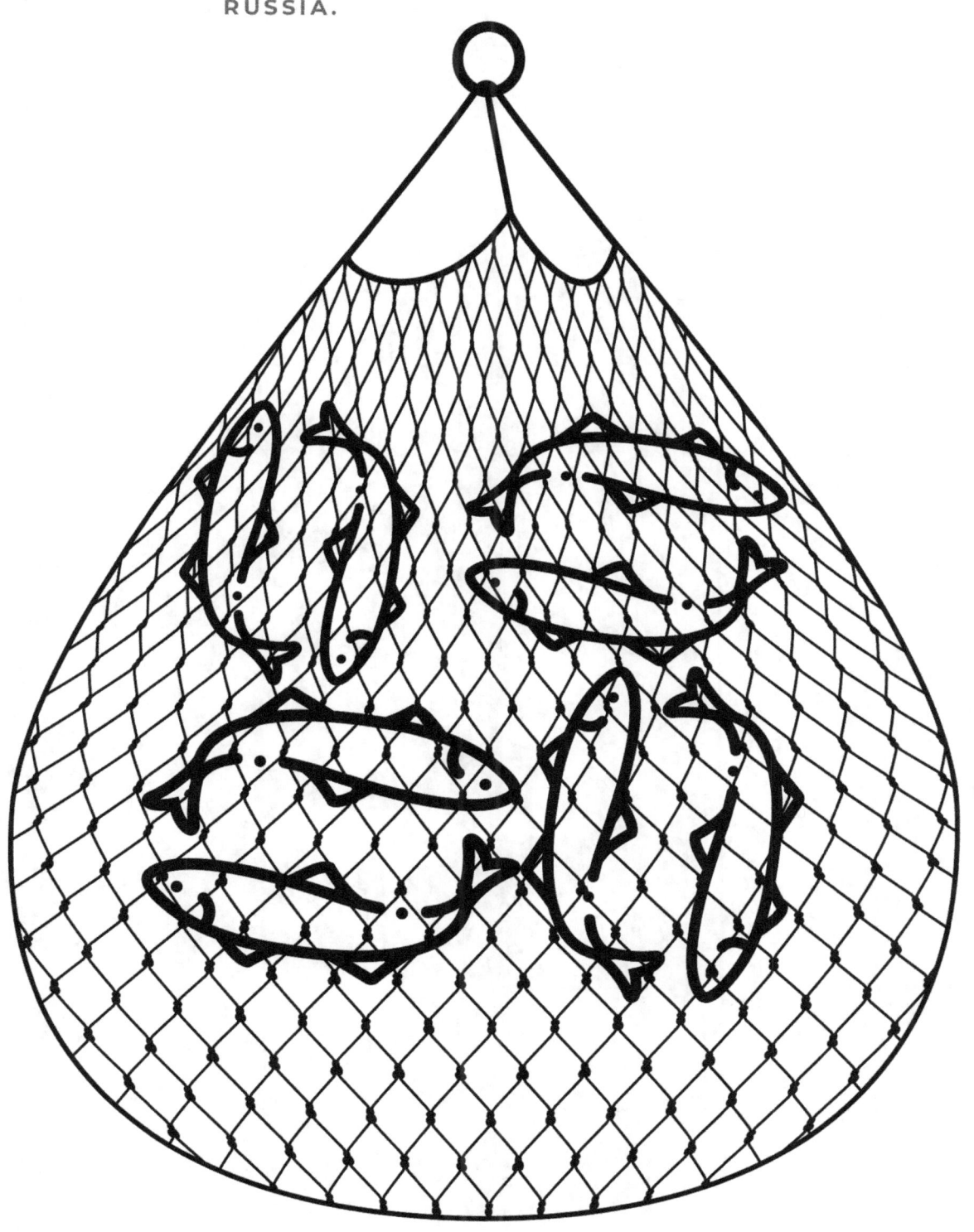

ARCTIC ECONOMICS

OIL AND GAS IN THE ARCTIC ARE MAJOR
ECONOMIC DRIVERS BECAUSE THE REGION
HOLDS LARGE UNDERGROUND RESERVES OF
THESE RESOURCES.

ARCTIC ECONOMICS

THE ARCTIC HOLDS VALUABLE MINERAL
RESOURCES, INCLUDING GOLD, IMPORTANT METALS
LIKE PALLADIUM, AND CRITICAL MATERIALS SUCH
AS RARE EARTH ELEMENTS (REES).

ARCTIC ECONOMICS

THE ARCTIC ECONOMY IS LIKELY TO GROW AS MELTING ICE OPENS NEW SHIPPING ROUTES AND ACCESS TO RESOURCES, BUT THIS GROWTH IS LIMITED BY ENVIRONMENTAL CONCERNS AND GLOBAL POLITICAL TENSIONS.

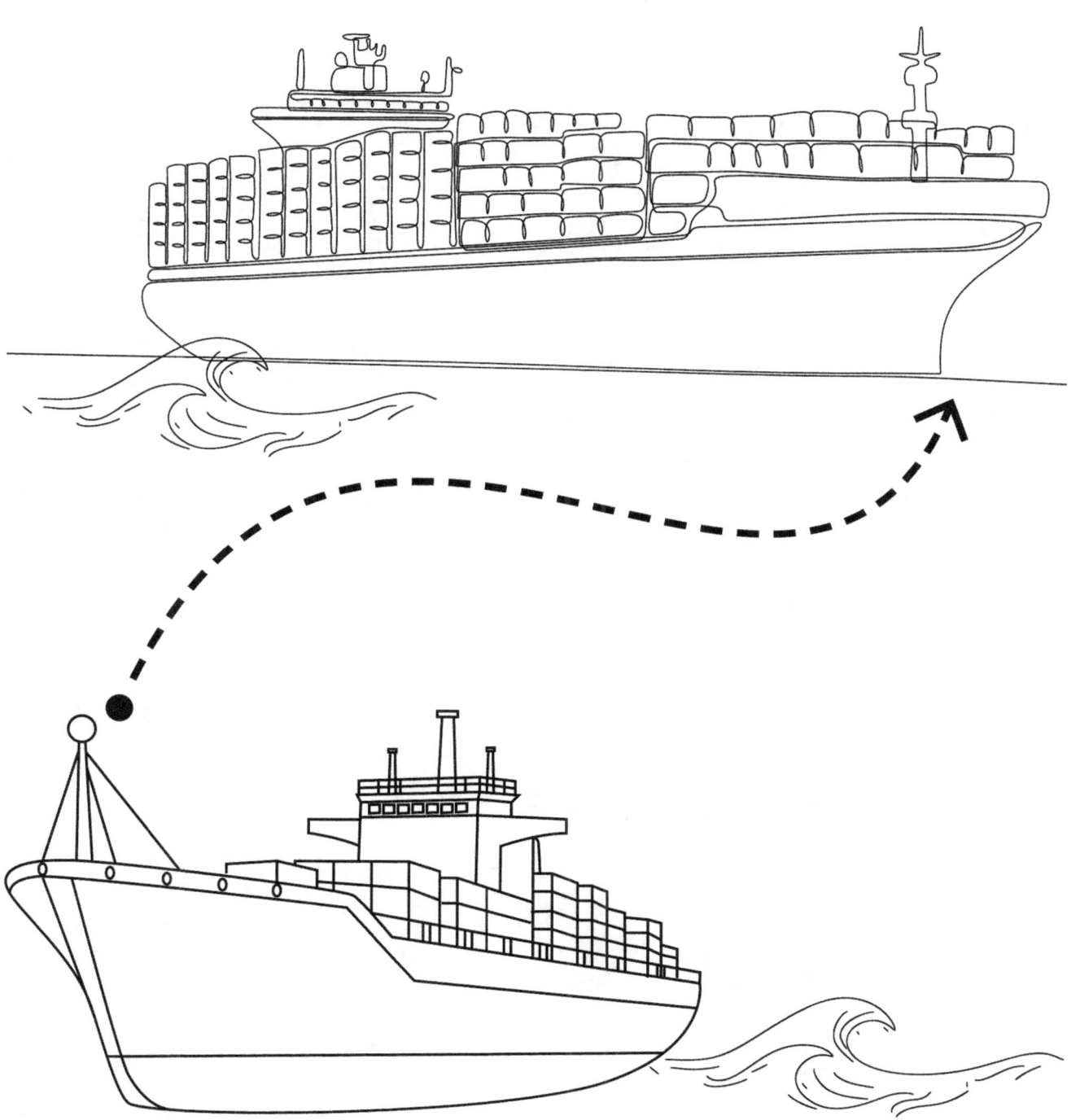

WOMEN IN TUNDRA ECOSYSTEMS

WOMEN IN THE TUNDRA ARE IMPORTANT TO
THEIR COMMUNITIES AND THE ENVIRONMENT,
ESPECIALLY AMONG INDIGENOUS GROUPS LIKE
THE INUIT AND SÁMI.

WOMEN IN TUNDRA ECOSYSTEMS

WOMEN TAKE CARE OF FAMILIES AND PLAY
KEY ROLES IN CULTURAL CEREMONIES,
EDUCATION, AND DECISION-MAKING.

WOMEN IN TUNDRA ECOSYSTEMS

IN MANY INDIGENOUS COMMUNITIES, WOMEN HELP GATHER PLANTS, BERRIES, AND FISH, AND SOMETIMES PARTICIPATE IN HUNTING.

WOMEN IN TUNDRA ECOSYSTEMS

WOMEN WORK IN FISHING, CRAFTS (LIKE SEWING TRADITIONAL CLOTHING), AND INCREASINGLY IN JOBS RELATED TO TOURISM OR LOCAL SERVICES.

WOMEN IN TUNDRA ECOSYSTEMS

LIVING IN HARSH CONDITIONS MEANS WOMEN FACE CHALLENGES LIKE LIMITED HEALTHCARE, EDUCATION ACCESS, AND THE IMPACTS OF CLIMATE CHANGE.

E
F P
T O Z
L P E D
P E C F D

WOMEN IN TUNDRA ECOSYSTEMS

WOMEN'S KNOWLEDGE AND WORK ARE ESSENTIAL FOR MAINTAINING CULTURAL TRADITIONS AND SUPPORTING SUSTAINABLE LIVING IN THE FRAGILE TUNDRA ENVIRONMENT.

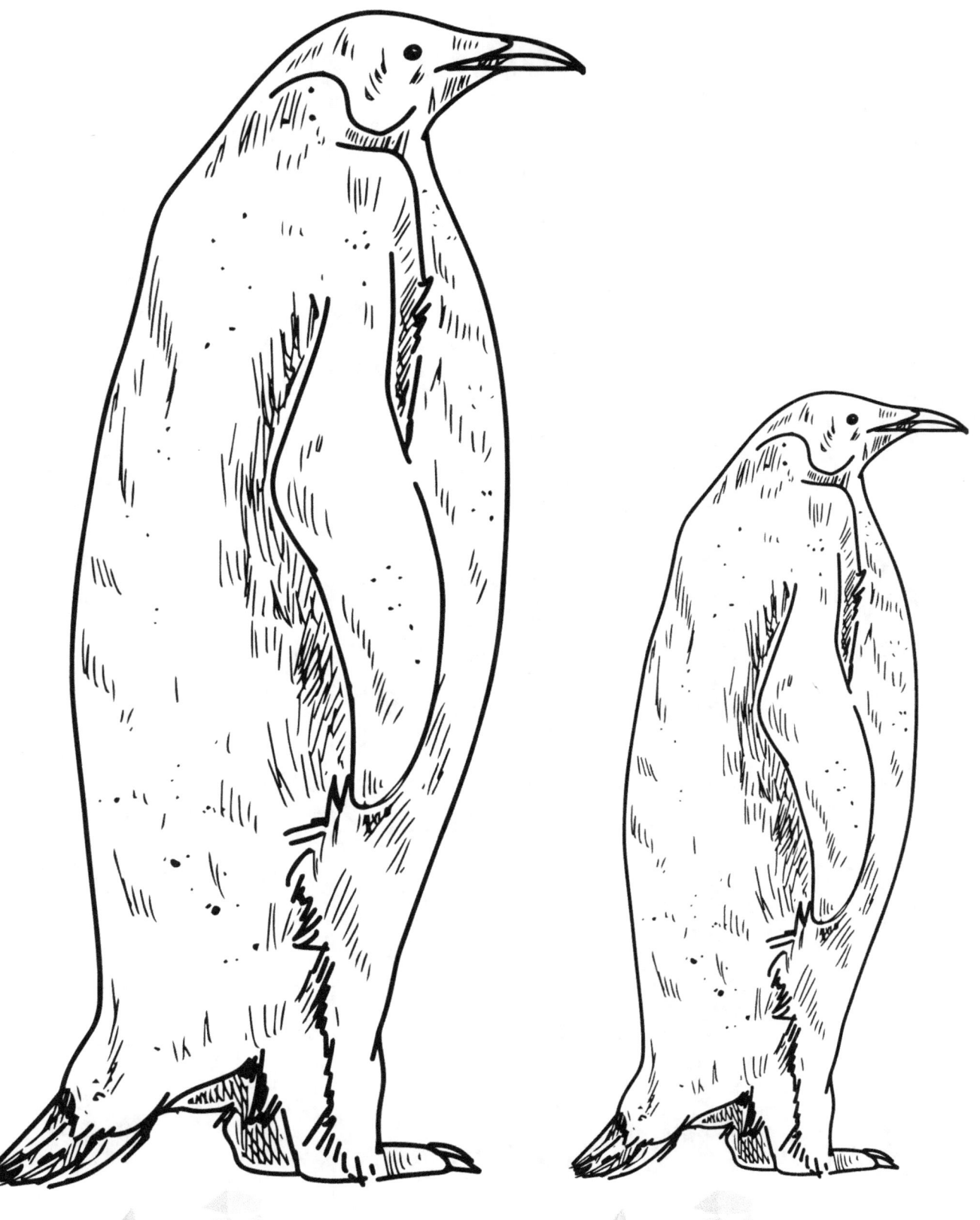

THE TUNDRA IS COLD AND TOUGH, BUT IT'S
ALSO QUIET, STRONG, AND REALLY BEAUTIFUL.

NOTES